7 plants and animals 8 sound

TEACHER'S GUIDE 7/8

Sylvia Jackson

Blackie

BLACKIE & SON LTD
Bishopbriggs
Glasgow G64 2NZ

Furnival House
14-18 High Holborn
London WC1 6BX

First published 1981
ISBN 0 216 90841 8

Printed in Great Britain by Thomson Litho Ltd, East Kilbride, Scotland

7

plants and animals

Book 7 can be divided into four main parts. The first part looks at the process of respiration. Respiration is of importance to both plants and animals and both types are investigated here. As a general rule all living things respire by using oxygen to burn up food and release energy. However, under certain circumstances, some living things can respire without oxygen. The first two activities in the book look at the way we respire and the gases involved. The third activity looks at the respiration of plants.

The second part of the book examines the growth of a plant from its seed. It looks at the type of conditions best for seeds to germinate and then follows the growth of a seed into a plant with a root system and a shoot with leaves.

Interrelated with this last section is the third part of the book which looks at how plants respond to certain stimuli such as light and gravity. Such growth movements are known as *tropisms*. In Book 2 pupils investigated *phototropism* (the movement towards light). In this book pupils go further and find out which colour of light gives the best response.

The fourth and last part of the book looks more closely at animals. First it looks in more detail at our senses, including those of taste and smell. It examines some of our reflexes and compares them with those of the earthworm. It then goes on to look at a number of small animals which children can collect themselves around their school.

INTRODUCTION

The introductory section simply outlines the main parts of the pupils' book and the type of experiments dealt with.

ACTIVITY 1
More about respiration

The main point to be made in Activity 1 is that the air we breathe in and the air we breathe out are significantly different. The air we breathe in contains more oxygen than the air we breathe out while the air we breathe out contains more carbon dioxide than the air we breathe in. Whatever method plants and animals use to take in and give out gases, the process of respiration within the living organism is the same: oxygen is used up and carbon dioxide is liberated. The experiment in Activity 1 depends upon the fact that a candle needs oxygen to burn; as a result, the air we breathe in will let the candle burn longer than the air we breathe out.

The first stage could be omitted and the candle stuck to the tin lid before the beginning of the experiment if you consider this part to be too messy or dangerous for your class.

In stage 4. when pupils blow through the tubing into the jar of water, air is first pushed out of the tubing. This is not air breathed out by the pupils. Therefore the air collected in the jar will contain a small amount of ordinary air. A more accurate result can be obtained by first blowing down the tubing to remove this air, pinching the bottom of the tube to close it and then putting it into the jar and proceeding as before.

In 7. the candle may be extinguished very quickly. This may be the result of carbon dioxide, which is denser than air, falling on to the candle and putting it out before any of the oxygen in the jar can get to the candle. One way of overcoming this problem is to attach the candle to a lid which will fit the jar. This is suggested later in Activity 3. Pupils could compare the results using the candle at the bottom of the jar and the candle attached to the lid. The diagram below shows the two arrangements.

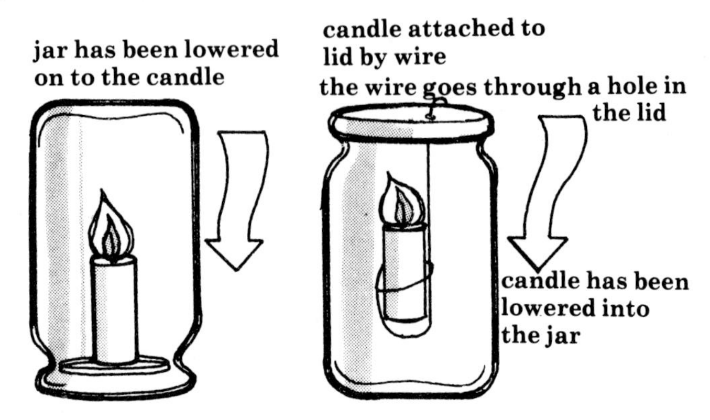

If you are worried that some air may be escaping from the jar through the hole where the wire goes through the lid, this can be made more airtight by putting vaseline around the hole. When a candle burns in air, using up oxygen, water is produced. Pupils therefore, may observe a misting on the inside of the jar, caused by the formation of water droplets.

In **Extra work** pupils may come across the role of *enzymes* in respiration. Enzymes are substances in the body which allow oxygen to burn up the food we eat (rather in the way that the candle uses oxygen as it burns) at low temperatures. Ask the children to imagine how uncomfortable it would feel if we had flames leaping up inside us.

ACTIVITY 2
More about oxygen

This activity aims to show that approximately one fifth of the air around us is oxygen. In **Burning a candle in the air**, at stage 3., the candle uses up the oxygen contained in the air of the jar and as it does, water rises up the jar to take its place. The method suggested in this activity of finding what fraction of the total length of the jar had water in it at the end of the experiment is only valid where the jam jar has the same circumference over its entire length.

Another method which should be used if the jar is an irregular shape is to find the volume of water which rises up the jar and find what fraction this is of the total volume of the jar. The jar can be filled up to the brim with water and then emptied into a measuring jug to find the total volume of the jar. The jar should then be filled up to the label mark with water and again emptied into a measuring jug to find its volume. The volume of oxygen used up is the difference between these two results. The fraction of oxygen in air can then be calculated by putting this figure over the total volume of the jar.

In **Extra work** pupils will discover that the remaining four fifths of the air is almost all made up of the gas, *nitrogen*. There is a very small amount of carbon dioxide and a group of other gases known as the *noble gases*.

ACTIVITY 3
Seeds and oxygen

This activity shows how seeds take in oxygen from the air as they grow. Germinating pea seeds are used. They should take about four days to begin sprouting. It is important that the seeds are used as soon as they begin sprouting. If not, leaves will quickly develop on the shoot enabling *photosynthesis* to take place (this is the process whereby plants make their own food). Photosynthesis is the reverse process to *respiration* where the gas, carbon dioxide, is taken in and the gas, oxygen, liberated. The results of the experiment will therefore be somewhat confused if photosynthesis proceeds alongside respiration. If the seeds are developing green parts on their shoots, one way of overcoming this problem is to put the jar in a dark cupboard so that photosynthesis is prevented.

At stage 3. it is important to make the jar as airtight as possible. Vaseline can be put around the edge of the lid holding the candle to make it more airtight.

In **Now try these questions**, pupils should have seen that the candle goes out much quicker in the jar containing the pea seeds than in the jar just containing air. Oxygen has been used up by the seeds in respiration so there is less oxygen for the candle which it needs to keep burning.

In **Extra work** pupils are asked to consider the difference between the living and the dead pea seeds. If the experiment is repeated using the same number of dead seeds, the candle will burn for about the same length of time as in the jar just full of air. This shows that only living seeds take in oxygen. One way of killing the germinating seeds is to boil them for about ten minutes in water and then to put them in bleach (e.g. Domestos) for about fifteen minutes. The seeds should be well rinsed with water before they are put in the jar with the lid on and left for the same amount of time as the living seeds.

ACTIVITY 4
Growing seeds

This activity can be introduced by asking pupils why it is that some seeds don't germinate. Why, for instance, do seeds bought in a packet not germinate if they are left in the packet? Why do some seeds not grow in the garden? Is it too cold? Are the seeds sown too deep? Are the seeds eaten by birds and other animals?

Activity 4 shows the importance of warmth, light and moisture in growing green healthy plants from seeds. It is important for pupils to check at regular intervals that those seeds in moist conditions are kept wet. In 8. pupils record how their seeds are growing, looking particularly at the colour and height of the shoot emerging.

Work in this activity could easily lead on to a comparison of the way seeds of different types grow. For example, do cress seeds grow quicker than pea seeds? In this type of experiment it is important to keep the conditions for growing both sets of seeds the same, otherwise the experiment would be unfair. You might want to make this point to the children. Another topic which could be looked at is the effect of overcrowding. Do ten cress seeds grow better spaced out or better very near together?

ACTIVITY 5
Looking at a broad bean growing

Activity 5 shows the pupils how well the broad bean seed is adapted for withstanding both adverse weather conditions and predators who may attempt to eat the seeds. Pupils use the information gained from Activity 4 to help their seeds germinate into green healthy plants by giving them warmth, moisture and light. Light is particularly important when the shoot is developing so that the green parts of the plant can produce their own food by the process of photosynthesis.

Before the seeds are put into the jam jar with the paper towel, pupils open up the broad bean seed to look at the creamy-coloured structures inside. These are called *cotyledons*. The broad bean seed has two of these and therefore is known as a *dicotyledon*. The cotyledons are the food store of the plant. They contain starch which turns iodine blue-black. The starch is gradually used up as the young plant (or embryo) grows into a mature plant and can make its own food. After about a week in the jar the bean should have developed a root (or *radicle*) which will have secondary roots growing out from its sides with root hairs attached and a shoot (or *plumule*) with leaves beginning to open.

At 7. and 8. pupils should have discovered that whatever direction the sprouting bean seed was put in the jar the root always grew downwards (with gravity) and the shoot upwards. By removing excess water from the jar, but remembering always to keep the paper towel moist, the jar can be turned upside down. The broad bean plants will respond so that the root turns to grow downwards and the shoot upwards. After a few days, when the jar is put the right way up again, it will look as if the root is growing upwards and the shoot downwards.

If time allows, pupils can contrast the growth of other seeds such as the french bean and the sunflower seed with the broad bean. Some seeds, like the two mentioned, push the seed up the jar (as if they were in the soil) before the shoot emerges out of the outer coat of the seed. This type of movement protects the shoot during growth. Another topic which can be investigated is the height to which different plants grow each day and pupils might make a graph of the results.

ACTIVITY 6
More about plants and light

Pupils have already seen in Book 2 how plants will grow towards the light. This activity shows how plants are attracted more towards certain colours of light than others. Bird seed grows very quickly and can be used instead of cress seeds if pupils want to look at another type of seed. If the seeds are grown in the dark they will develop into tall plants very quickly. Bird seed which has been soaked can finish its growth in a flower pot in soil.

ACTIVITY 7
More about our senses

In the first part of this activity pupils will find difficulty in distinguishing between apple and onion unless they are able to smell as well as taste. If pupils have difficulty in smelling the food when it is placed on their tongue, they should be allowed to smell it first near their noses.

The second part of the activity is concerned with how different parts of the tongue are sensitive to different tastes. Details of the four solutions used are given in the materials list at the end of this section. The four main tastes which the tongue is sensitive to, are sweet, sour, bitter and salt. The diagram opposite shows which part of the tongue is sensitive to each taste.

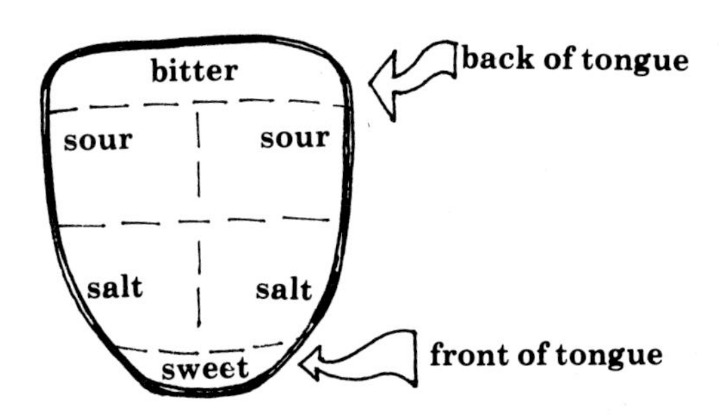

The final part of Activity 7 looks at our sense of balance. The parts of our ear which are responsible for our sense of balance are called the *semicircular canals.* In Book 8, *Sound*, page 31, there is a picture of the ear showing the position of the semicircular canals. In this teachers' guide, page 25, there is more information about this part of the ear. The semicircular canals are contained in the inner ear with the cochlea. There are three semicircular canals which lie approximately at right angles to one another. They are filled with a fluid and it is the effect of the movement of this fluid against the sensitive areas in the semicircular canals (that are responsible for relaying messages to the brain) that gives rise to our sense of balance. The reason the pupil still thinks he is moving even when the chair has stopped revolving is because of the continuing movement of this fluid. This part of the ear also tries to rectify the situation when the chair stops moving. This gives rise to the pupil feeling as if he is spinning around in the opposite direction.

ACTIVITY 8
What is a reflex?

In the first experiment in Activity 8 (**Your reflexes**) pupils study two reflexes, the leg jerk and the change in the size of the pupil of the eye when we look at a bright light. The first of these should be done as a demonstration. In the second a *dimmed* torch should be used as a safety measure, i.e. a torch covered with tissue paper. This reflex is even more pronounced if the pupil undertaking the experiment first closes his eyes, then opens them, and then is subjected to the light. Other reflexes include coughing, jumping from fright and laughing when tickled. Reflexes are important to humans because many of them prevent the body being injured; a quick mechanical response takes the body away from harm.

Whereas much of our behaviour is an expression of learned responses, less sophisticated animals, such as the earthworm, depend more on their reflexes. These reflexes are automatic and unlearned. In the second part of this activity (**Looking at the earthworm**) pupils see how simple touch and the smell of vinegar can cause a simple reflex action in the earthworm. The earthworm jumps back in a similar way to how we might if we touched a hot object by accident. Care must be taken not to touch the earthworm with the vinegar as this can seriously injure or even kill the earthworm.

In **Extra work** we look at the brain. In man the brain varies in weight from 1000 g to 1600 g. The brain is that part of the nervous system enclosed in the skull and is made up of three main parts: the *forebrain,* the *midbrain* and the *hindbrain.* The different functions which the different parts carry out are shown in the diagram in the pupils' book. Pupils can obtain pictures of the brains of other animals such as fish, amphibians, reptiles and birds to compare the sizes of the three parts compared with man.

ACTIVITY 9
Looking at small animals

In Activity 9 children come into contact with a number of small animals, some of which will be familiar to them already. However, they may not have observed these animals in any detail before, nor may they have read much to supplement their observations. As well as drawing pictures of the animals they look at, pupils might also make models.

The second part of Activity 9 (**Finding out more about the animals**) looks in more detail at the habitat of the animals collected. You could ask the children to imagine what it would be like living under a stone: Would it be light or dark? Ask them to feel the soil: is it wet or dry? Can they think of any reasons why these animals should live in this type of place? The experiment described in the pupils' book is one way of testing whether these animals prefer light or dark conditions. If they have time, pupils can also test whether their animals prefer damp or dry soil by using another dish with dry soil at the left-hand side and damp soil at the right-hand side. Again, the animals would be put in the centre of the tray and we would observe which way they moved. One way of drying the soil is to heat it up in an old saucepan or tin lid.

Materials List

ACTIVITY 1

Tin lid (the circumference should be larger than the jar), vaseline, sink or bucket of water, jam jar, clean piece of tubing or straw, short candle and tin lid (the lid's circumference should be smaller than the jar), matches, watch with a second hand or stop watch.

ACTIVITY 2

(Burning a candle in the air) Sink or bucket of water, short candle and tin lid (as used in Activity 1), matches, labels, jam jar (larger than tin lid).
(Finding the amount of oxygen in the air) Jam jar with label mark from the first experiment in Activity 2, ruler.

ACTIVITY 3

Pea seeds, paper towel, tray, water, cottonwool, vaseline, screw top jar and lid, short candle fastened to lid (which will also fit jar), matches, watch.

ACTIVITY 4

Cottonwool, water, 8 small jars, e.g. baby food jars, cress seeds.

ACTIVITY 5

5 broad bean seeds, magnifying lens,* paper towels, water, tray, spoon, iodine, jam jar.

ACTIVITY 6

Paper towel, 3 tin lids, water, cress seeds, 3 shoe boxes, scissors, cellophane (red, yellow and blue – each piece must be sufficiently large to fit over one end of the shoe box), sellotape.

ACTIVITY 7

(Does smell help our sense of taste?) Blindfold, 20 pieces of onion, 20 pieces of apple, 2 clean spoons, beaker of water.

(Can different parts of the tongue pick out different tastes?) Blindfold, 4 solutions – salty (salt dissolved in water), sweet (sugar dissolved in water), sour (lemon juice), bitter (cold tea), spoon, coloured pens.
(Your sense of balance) Blindfold, revolving chair.

ACTIVITY 8

(Your reflexes) Chair, torch (dimmed with a piece of tissue paper sellotaped over it).
(Looking at the earthworm) Earthworm, stick (e.g. lollipop stick), cotton-wool, vinegar.

ACTIVITY 9

(Finding and looking) Large dish or tray* (approx. 30 cm × 30 cm × 10 cm), small dish (e.g. margarine tub), two pieces of cardboard (to fit over the two dishes), spade or trowel, spoon, brush, ruler, magnifying lens.
(Finding out more about the animals) Large dish of soil and stones with large piece of cardboard from first part of the activity, large dish (empty), piece of cardboard to fit over empty tray, spoon, brush, scissors.

* Suppliers include Griffin & George, Philip Harris and Osmiroid.

8

sound

Book 8 can be divided into six main parts. The book begins by examining the different ways that sounds can be made and shows that all of these methods involve vibration. The second part highlights the wave nature of sound, showing how sounds, like light (discussed in Book 4) can be reflected or made to bounce back from an object. Pupils then examine how sound travels through solids and liquids as well as gases such as air. The fourth and fifth parts of the book are concerned with how the changes in the loudness and pitch of a sound are brought about. The last part looks at our own ears and also those of other animals.

INTRODUCTION

The introductory section begins by listing some of the everyday sounds around us. This includes the children drawing up their own list of sounds they hear around the school. This work could be undertaken in groups with each group allocated to a different part of the school.

ACTIVITY 1
More about making sound

One way of introducing this topic would be to ask pupils to feel their own vocal cords. If they place their fingers lightly on the throat just below the chin and sing a low note they should be able to feel the vocal cords vibrating.

Activity 1 shows three ways of making a sound: by banging (using a tuning fork and drum); by plucking (using the ruler and rubber band); and by blowing (using the milk bottle). In each case the pupils should see that a sound is made when something (the tuning fork, drum, ruler, rubber band or air in the milk bottle) vibrates.

In **Extra work** (page 4) musical instruments which fit into the three categories include:
1. Different types of drums, chime bars, triangles (all giving a sound when they are hit).

2. Violin, harp, double bass (all giving a sound when they are plucked).
3. Flute, trumpet, recorder (all giving a sound when they are blown).

It would certainly add to the interest of the topic if a selection of musical instruments could be looked at and tried by the children. Those children who play a musical instrument could be asked to give a talk about how it works. This would also be useful later in Activities 4 and 7 where the loudness and pitch of sound are investigated.

ACTIVITY 2
Using our ears

In the first experiment (**Guessing the direction of sounds**) pupils may find difficulty in locating the exact direction from which the clap is made. Sounds which are most clearly heard are those in line with our ears and this fact may be brought up by the pupils doing the experiment. A number of groups could undertake the experiment and compare their results. You might ask if it is fair to compare these results. Was all the clapping of the same loudness? Was the pupil in the centre of each group the same distance away from the pupils standing in the circle?

The second experiment (**Is one ear better than the other?**) raises similar questions about the fairness of comparing results; most importantly is just one result for each pupil sufficient for comparison? You may suggest that each pupil should have five attempts and then the average could be found as in the first experiment. Also, does it make any difference to the results if some of the pupils know which side the watch is? You might suggest that the experiment could be improved if a pupil was to stand and move in the same way at each side of the pupil blindfolded, so that it would be impossible for the person tested to tell where the watch is, except by hearing.

Some pupils will perform slightly better than others in these experiments. However, some children may have more difficulty because of temporary deafness caused by a cold or waxing up of the ears, or because of a more permanent defect. Pupils will look at the ears in more detail in Activity 9.

ACTIVITY 3
Can we make sounds bounce back?

Sound travels in waves like light. It therefore shows the same properties as light. One of these is that sound, like light, can be reflected. When sound waves hit a flat surface they will bounce back away from the surface. This is shown in the diagram below.

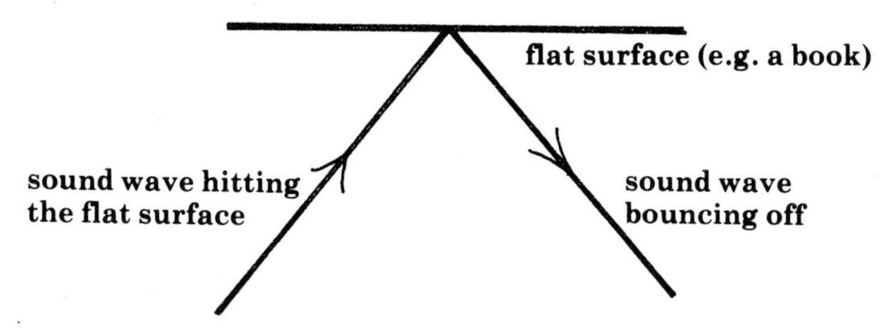

If a line is drawn at right angles to the flat surface, then the angle that the in-going sound wave makes with this line is equal to the angle the out-going (or reflected) sound wave makes with it. This is shown in the diagram below.

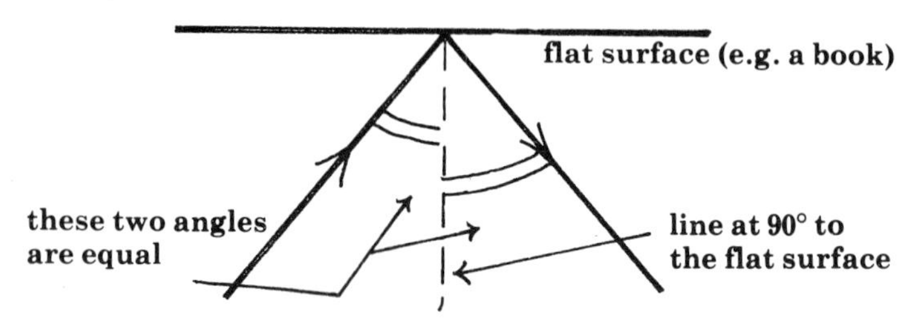

Therefore pupils will find the best results for hearing the reflected sound when the two cardboard tubes are placed in positions giving equal angles to an imaginary line drawn at 90° to the surface (in their case the book).

The cardboard tube holding the watch channels the sound waves towards the surface of the book. When the sound waves hit the book, they are reflected back, and channelled along the second cardboard tube to the pupil's ear. Sound travels through solids. Therefore, if the cardboard tubes and book are touching, pupils may hear the watch, not because of reflection, but because the sound has passed from the watch through the solid tubes and book to their ear. Even if the objects are not touching, some sound will travel from the watch through the desk on which the cardboard tubes sit.

At 4., when the listening tube is removed, it becomes very difficult to hear the ticking watch. Although the sound bounces back from the book it spreads out into the air and is not channelled back to the ear. At 5., when the book is removed, there is no surface for the sound waves to bounce back from and therefore the loudness of the ticking is reduced. At 6. the ticking is again reduced because the source of the sound is less strong. Without the cardboard tube to channel the sound to the book, less sound will hit the book and bounce off in the direction of the pupil's ear.

In **Other questions** pupils are asked to use their library books to find out more about echoes. An echo occurs when a sound is reflected back from a solid object like a wall or cliff so that two sounds can be heard – the original sound (e.g. a clap or a shout) and its echo. Question 2. looks at echo-sounding. This is a method of calculating distance by using the time it takes for the echo of a sound to travel back. It is used by ships at sea to determine the depth of the sea bed. A sound is sent down from the ship to the sea bed. When it hits the sea bed it bounces back and the echo is picked up on board ship. Knowing how fast sound travels in the water and the time interval between the initial sound made and its echo, the depth of the sea bed can be calculated. Question 3. is about how echoes can be reduced in theatres and cinemas. A certain amount of echo (sometimes called *reverberation*) is needed as this amplifies and strengthens the original sound. However, if the reverberations last too long and are too strong, they begin to interfere with other sounds which follow. Smooth surfaces enhance the reverberations and are therefore avoided. Instead, materials which absorb sound are used, e.g. curtains, carpets, etc.

Some pupils may know of places near at hand where a good echo can be heard. Possibly this could be visited by the class. You need to be about 20 metres from the wall (or reflecting surface you are using) to hear a distinct echo.

ACTIVITY 4
Loud and soft sounds

Activity 4 begins by showing that the loudness of a sound can depend upon how hard you bang, pluck or blow to make the sound.

In the first experiment (**Making soft sounds louder**) pupils discover how different cardboard shapes alter the loudness of sound to different degrees. With the wide end of the cardboard cone nearest to the ear the loudest ticking is heard. Using the cone the other way round makes the ticking very soft. When the watch is placed inside the cardboard tube the loudness is somewhere between that of the two cones. Care is needed to make sure that the watch does not fall out of the end of the cardboard shape. A good size for the cone is 10 cm diameter at the wide end and 4 cm at the other end. The tube should then be 4 cm diameter. Both should be about 20 cm long.

In the second experiment (**Making loud sounds softer**) pupils find out how fabrics such as those that are used for blankets, carpets and curtains absorb sounds, making them softer. As the clock is lowered into the box lined with these materials the ticking appears much fainter than it did without the materials in the box. Using the alarm of a clock gives an even more impressive performance.

In **Extra work** (page 15) both questions are concerned with the ways sound is absorbed by materials. Drawing the curtains across in one's home can reduce the sound heard from a television. In the home and at school sound is not only reduced when certain types of floor covering are used, such as carpeting, but also by the type of soles on shoes, by the types of sound-absorbing material put underneath the legs of tables and chairs, and by the type of materials used for walls and ceilings.

ACTIVITY 5
How does sound travel?

There are two main types of waves called *transverse* and *longitudinal* waves. Light waves are an example of transverse waves. This type of wave action can easily be shown by moving a pencil up and down in a dish of water (as in 1.), or by moving a skipping rope or slinky from side to side (as in 2. and 3.). As the water, skipping rope or slinky vibrates, its motion is perpendicular to the direction in which the waves are moving. The motion of the water can be easily shown by the action of a cork bobbing up and down on the water. A piece of coloured wool tied to the middle of the skipping rope or slinky shows that they move from side to side as the waves move along.

The second type of waves are called longitudinal waves. Sound waves are longitudinal waves and 4. tries to show the difference between this type of waves and transverse waves. This time instead of the slinky moving from side to side, perpendicular to the motion of the waves, it vibrates in the same direction as the motion of the waves. This is shown in the diagrams overleaf.

While it is not necessary at this stage to use the terms transverse waves and longitudinal waves, most pupils should be able to see a difference between the two types of waves, especially in 3. and 4. The first two parts of this activity can be used as a general introduction to waves showing how they are produced. Some pupils might be able to relate the type of waves produced with the water and skipping rope to the two types seen later with the slinky.

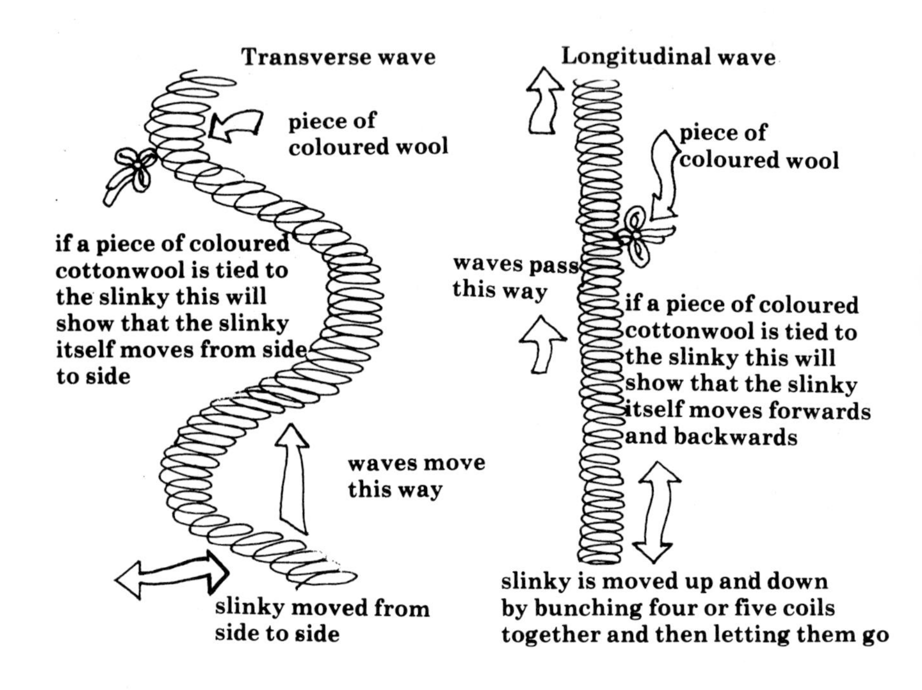

ACTIVITY 6
More about sound travelling

Activity 6 shows how sound will travel through solids and liquids as well as through the air. In 2. the squeaky toy still makes a sound when it is under the water. Sound travels easily through the desk in 3., so that even a very gentle tapping sound should be heard.

In 4., 5. and 6. pupils construct and try out a simple telephone. The sound travels along the string from one tin can to the other. The string telephone can be tried out of doors using a long piece of string (it can be 30 metres long) to show even more convincingly how sounds can be heard better with the telephone. In 6. pupils will hear their friends speak just as well as if they were using the telephone in the room. In 7. and 8. the sound of the spoons hitting each other passes very effectively up the string to the pupils' ears.

ACTIVITY 7
Pitch

This activity can be easily compared with Activity 4. Whereas in Activity 4 the rubber band was plucked harder to make the sound louder, this time the rubber band is first held loosely and then tighter giving two sounds of different pitch.

In 1. and 2. the shorter the length of ruler hanging over the bench the higher the pitch of the note. In 3. and 4. the more tightly stretched the rubber band the higher the pitch of the note. Stretching the rubber band over a matchbox or an open tin can will make the sounds easier to hear. The pitch of the sound made by the rubber band can also be changed by using a different length of band; using a shorter length produces a higher pitched note.

In 5. to 8. pupils discover how bottles containing different amounts of water produce notes of different pitch. In 6., where pupils are blowing across the mouth of the bottle, the smaller the amount of air in the bottle the higher the pitch of the note produced. However in 7., where the bottles are being hit with a spoon or hard stick, the bottles must be arranged in the opposite direction, as the bottle with most air in (or least water in) will now give the highest note. If the eight bottles used in 8. have been labelled A to H the following letters give the tune to 'The First Nowell'.

C B A B C D E F G H G F E F G H G F E F G H E D C

Repeat above once more

C B A B C D E H G F F E H G F E F G H E D C

In **Extra work** there is no mention of any stringed instrument as this is dealt with in the next activity. In instruments such as the trombone and organ the pitch of the note depends upon the length of the column of air which is vibrating. In the trombone this is done by changing the length of the tube. In the organ individual tubes or pipes of different lengths are used. Chime bars consist of bars of different lengths which produce notes of different pitch when they are hit.

ACTIVITY 8
More about pitch

Activity 8 extends the work already undertaken on pitch. At the beginning of the activity pupils should be able to use their knowledge gained from Activity 7 to guess that tightening the string by turning the screw eye will cause the string to vibrate quicker and so give a higher pitched note. If pupils have difficulty screwing in the screw eye (at stages 1. and 2.), you can help them by making a small hole in the wood with the corner of a screwdriver. In stage 6. the screw eye may be stiff to turn. An easier way is to put the metal end of a screwdriver through the hole and pull the handle of the screwdriver round, so turning the screw eye. Once the screw eye has been turned forwards and backwards a few times it will be easy for the pupils to turn it themselves. String has been suggested for this experiment rather than wire because it will be considerably safer if it breaks when stretched.

In **Extra work** weights can be used to tighten the string by using the type of apparatus shown below.

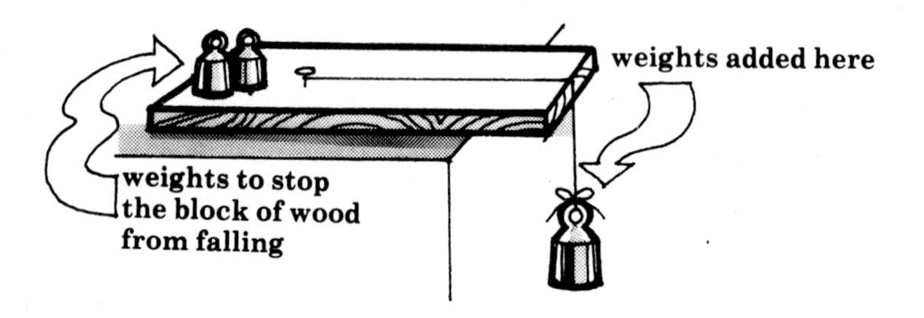

As more weights are added to the string it becomes tighter. When the string is plucked it vibrates quicker causing a higher pitched note to be produced.

Later in the activity (stages 8. and 9.) pupils may find that they need a friend to hold the left-hand side ruler in place to prevent it falling down. Moving both the ruler and the finger down the string effectively shortens it. The string vibrates quicker producing a higher pitched note.

The last part of Activity 8 (stages 14. to 18.) shows pupils how to construct and tune their own three-stringed musical instrument.

In **Extra work** pupils look more closely at the violin, examining how the bridge lifts the strings up so that they can be more easily played using a bow. You may wish to compare the instrument made by the children with the violin. The screw eyes turn to tighten the strings just as the pegs on the violin can be adjusted. As the pupils discovered in stages 13. and 18., so the violinist changes the pitch of notes by pressing his finger down at different positions along the strings.

ACTIVITY 9
Ears

The ear consists of three main parts called the outer, middle and inner ears. The *pinna* and *tube leading to the eardrum* (called the external auditory canal) make up the outer ear. The middle ear is made up of three bones: the *hammer*, the *anvil* and the *stirrup*. These are so named because of their shape. The *eustachian tube* is a narrow canal leading from the middle ear to the throat. It keeps the pressure on both sides of the eardrum equal. The inner ear consists of two parts: the *cochlea* (so named from the Latin meaning 'snail's shell') and the *semicircular canals*. The semicircular canals are responsible for our sense of balance and have already been discussed in Book 7.

In **Now try these questions** the first question concerns the function of the eustachian tube. If the eustachian tube becomes blocked as it can when we have a cold the pressure is no longer equal on each side of the drum and as a result hearing becomes more difficult. Questions 2. and 3. look at the function of wax. It is responsible for keeping the eardrum lubricated. However, the secretion of too much wax can cause the tube leading to the eardrum to become blocked, so impairing our hearing. It may then be necessary to have the ears syringed by a doctor or nurse. Question 4. is about the possible dangers of listening to noises which are too loud. This can lead to pain in the ear and in some cases long-term damage to the inner ear.

Question 5. looks at the question of deafness. We have already referred to temporary deafness caused by a cold or the waxing up of the ears. Also the middle ear can become damaged. For example, the bones of the middle ear cease to function properly because tissue obstructs their movement. In the inner ear the cochlea itself can become diseased or damaged. The presence of tumours and abscesses in parts of the ear will also give rise to deafness.

Question 6. looks at ultrasonic sound. Humans hear sounds which have a frequency of between 20 and 20,000 cycles per second. Sounds which have a frequency of greater than 20,000 cycles per second are called *ultrasonic sounds*.

The second part of Activity 9 looks at the way other animals hear. Unlike man, where the pinna of the ear can have a limited function, the rabbit uses it for catching and directing sound waves much more effectively. The ears of a rabbit are constantly turning in different directions. The ears can be moved independently so that while one is turned to the front the other may be turned to the back. When a sound is heard, both ears are turned in that direction. The bat uses its voice and its ears together to locate obstacles and food. It gives out high-pitched squeaks (in the range of ultrasonics so that they are not heard by humans) which, if they hit anything, send back an echo which the bat picks up with its ears. The owl is more precise at locating the direction of a sound than even the rabbit. The barn owl is particularly sensitive and can hunt with deadly accuracy in complete darkness using just its ears. In addition to being good direction finders the barn owl's ears are sensitive to sounds above the range of their own voices. This is different from most other birds. Thus the owl, unlike most birds, is able to pick up the high notes of mice squeaking.

The grasshopper has ears on its body. The short-horned variety has them on each side near to where the abdomen is jointed to the thorax. The long-horned grasshopper has ears on its legs. The ears of both types of grasshopper work in a similar way to our own, although they are rather simpler. Fish like the salmon have no outward sign of an ear, although they possess that part of the ear we refer to as the inner ear. It seems that a small part of the inner ear may be concerned with hearing, the larger part being concerned with balance. However, in addition to the ear, fish such as the salmon, possess another means of detecting vibrations in the water around them. They use a set of sense organs running along the side of the body, known as the lateral line. By helping salmon to detect movements in water this line tells the fish whether another fish or obstacle is near.

Materials List

ACTIVITY 1

Ruler, tuning fork, rubber or cork, dish, jug of water, thread, rubber band, table tennis ball, rice or small pieces of paper, milk bottle.

ACTIVITY 2

(Guessing the direction of sounds) Blindfold.
(Is one ear better than the other?) Watch, ruler or tape measure.

ACTIVITY 3

Two cardboard tubes (e.g. from inside a roll of cooking foil) about 20 cm long, book, watch.

ACTIVITY 4

(Making soft sounds louder) Cardboard tube and cone (about 20 cm long), watch, ruler or tape measure.
(Making loud sounds softer) Large box (about 30 cm square and 60 cm deep), alarm clock, string (at least 100 cm), scissors, material for lining box (e.g. a blanket, curtain and a piece of carpet), ruler or tape measure.

ACTIVITY 5

Dish, jug of water, pencil, skipping rope, desk or chair, slinky.*
Extras: cork, piece of coloured wool.

ACTIVITY 6

Squeaky toy, plastic bag, string, scissors, sink or bucket of water, desk, 2 tin cans (each with one hole in the bottom).

ACTIVITY 7

Ruler, rubber band, 8 identical bottles, water, spoon, labels.

ACTIVITY 8 Block of wood (approx. 40 cm × 20 cm), string (approx. 2 m), scissors, 6 screw eyes, screwdriver, 2 wood rulers.

ACTIVITY 9 Library books.

* Suppliers include Griffin and George, and Philip Harris.